Fox Poems

Compiled by John Foster

D1533563

OXFORD

Oxford University Press, Walton Street, Oxford OX2 6DP

Oxford New York Toronto
Delhi Bombay Calcutta Madras Karachi
Petaling Jaya Singapore Hong Kong Tokyo
Nairobi Dar es Salaam Cape Town
Melbourne Auckland

and associated companies in
Berlin Ibadan

Oxford is a trade mark of Oxford University Press

© Oxford University Press 1991
Printed in Hong Kong

A CIP catalogue record for this book is available from the British
Library.

Acknowledgements
The Editor and Publisher wish to thank the following who have
kindly given permission for the use of copyright materials:

Stanley Cook for 'The fox who came to town' © 1990 Stanley Cook;
Julie Holder for 'The Corn Scratch Kwa Kwa hen and The Fox' ©
1990 Julie Holder; Irene Rawnsley for 'Foxes Sleeping' and
'Chicken and Fox' both © 1990 Irene Rawnsley; Jill Townsend for
'Red Fox' © Jill Townsend; Raymond Wilson for 'The Fox and The
Crow' © 1990 Raymond Wilson.

Although every effort has been made to contact the owners of
copyright material, a few have been impossible to trace, but if they
contact the Publisher correct acknowledgement will be made in
future editions.

Chicken and Fox

Bag-of-Grain
The farmer's chicken
Found a clock;
She liked its ticking.

Came a goose
To ask the time;
Chicken told her
Half past nine.

Came the farmer's
Splendid cock;
Chicken told him
Twelve o'clock.

Came a knock
At half past four;
Fly the Fox
Stood at the door.

Bag-of-Grain
Was full of fear;
Said to him, 'Sir
Your clock's not here;

I lent it out
To Lardy Hen
Who's fast asleep
In the farmyard pen.'

2

Five o'clock
From his window high
The farmer spotted
Foxy Fly.

Took from its case
A great big gun;
Soon Fly the Fox
Was on the run.

Now Bag-of-Grain
Has a tale to tell;
Seven o'clock
And all is well!

Irene Rawnsley

3

The Fox and the Crow

A farmer's wife threw out some cheese,
    and before you could count three,
A crow swooped down and carried it off
    to the top branch of a tree.

All this was seen by a hungry fox,
    who called up to the crow:
'How *very* beautiful you are!
    Has no one told you so?'

Now since the crow still held the cheese
    quite firmly in his beak,
He gave a nod to the fox below,
    but didn't dare to speak.

'Not only are you beautiful,'
    the fox said, 'but I've heard
your voice is lovelier than the voice
    of any other bird!'

The crow, puffed up by all of this,
    smiled down at the fox below,
But his beak still firmly held the cheese
    and did not let it go.

'If I could hear your song,' said the fox,
    'I'd soon be able to tell
if it's true that even the nightingale
    cannot sing half as well!'

At once, the crow broke into song –
    a single, ugly 'Caw' –
And the cheese fell from his open mouth
    on to the forest floor.

The fox, quick as lightning, snapped it up
    and laughed to think such a prize
Could be won from a crow stupid enough
    to fall for a pack of lies!

Raymond Wilson   5

The Rabbit and the Fox

A rabbit came hopping, hopping,
Hopping along in the park.
'I've just been shopping, shopping,
I must be home before dark.'

A fox came stalking, stalking,
Stalking from under a tree.
'Where are you walking, walking?
Why don't you walk with me?'

The rabbit went hopping, hopping,
Hopping away from the tree.
'I've just been shopping, shopping,
I must be home for my tea.'

'Come with me, bunny, bunny –
Bunny, you come with me;
I'll give you some honey, honey,
I'll give you some honey for tea.'

'I can't be stopping, stopping,
I'm far too busy today' –
And the rabbit went hopping, hopping,
Hopping away and away.

Clive Sansom

7

The fox who came to town

When the fox decided to move
He was far too knowing and clever
To travel by a busy road
And risk being hit or run over.

He went along the railway line
Where trains once ran to town,
Where the track was taken up
And the signals taken down.

He passed the empty stations:
No passengers were waiting there,
The booking halls were closed
And no one asked for his fare.

He reached the middle of town
With its noisy, busy streets
And quietly stole about,
Looking for something to eat.

Stanley Cook

8

The Corn Scratch Kwa Kwa Hen and the Fox

And the Corn Scratch Kwa Kwa Hen
Heard the grumbling rumbling belly
Of the Slink Back Brush Tail Fox
A whole field away.

And she said to her sisters in the henhouse,
'Sisters, that Slink Back Brush Tail Fox
Will come and here's what we must do,'
And she whispered in their sharp sharp ears, 'kwa kwa.'

And when that Slink Back Brush Tail Fox
Came over the field at night,
She heard his paw slide on a leaf,
And the Corn Scratch Kwa Kwa Hen and her sisters
Opened their beaks and —

'KWA!'
The moon jumped
And the Chooky Chook Chicks
Hid under the straw and giggled,
It was the **LOUDEST KWA** in the world.

And the Log Dog and the Scat Cat
And the Brat Rat and the House Mouse
And the Don't Harm Her Farmer
And his Life Wife and their Shorter Daughter
And their One Son came running,

On their slip slop, flip flop,
Scatter clatter, slick flick, tickly feet
And they opened their mouths and shouted –

'FOX!'
And it was the **LOUDEST NAME** in the world.
And the Slink Back Brush Tail Fox
Ran over the fields and far away
And hid in a hole with his grumbling rumbling belly.

And the Corn Scratch Kwa Kwa Hen
Tucked the Chooky Chook Chicks under her feathers
And said 'kwa,'
And it was the softest kwa in the world.

Julie Holder

Foxes Sleeping

Deep in her earth
within the woods
the mother fox sleeps
with four young cubs.

Outside the den are
feathers and bones.
Last night the foxes
dined at home.

Chicken and pigeon
under the moon:
fox-father will come
with more food soon.

Tracks and scent
mark the way he'll take.
He'll come with a rabbit
before they're awake.

Irene Rawnsley

Red Fox

Long and lean, he is a shadow
thinking itself between trees.
A red flick of ear or leaf —
then he waits, listening.
A dog's bark cuts
the sharp night air.
He thinks himself away.

Jill Townsend